Introduction to Supramolecular Chemistry and Optical Sensing Technology

Shampa Chakraborty
Moumi Mandal
Sadhana Rayalu

ELIVA PRESS

ELIVA PRESS

Shampa Chakraborty

Moumi Mandal

Sadhana Rayalu

Supramolecular Chemistry is one of the most fashionable and budding areas of chemistry. Supramolecular chemistry studied the area of protein folding, host-guest chemistry, interlocked molecular architectures, and dynamic non-covalent chemistry. Supramolecular Chemistry or Molecular recognition deals with the detection of ionic or neutral guests by specific interactions with the host molecule. One of the most useful application for supramolecular chemistry is chemical sensing. Selective chemical sensors are very useful for the monitoring of lot of samples and the methods are convenient and cheap, easily operable and require no pre-treatment. An ion sensor should be selective for a particular host material but has poor sensitivity to others. The sensing technology is one of the main detection tools for modern chemistry as this will give prompt and accurate results without any pretreatment. The beginning of supramolecular chemistry has been considered since 1987 when Charles Pederson won the Nobel prize for synthesizing cyclic polyethylene glycol (1967) (crown ether) – a metal ion scavengers. Over the five decades, Supramolecular Chemistry evolutes and creates its applications in many newer branches. As a result, again in 2016, the Nobel Prize for chemistry comes in the same field introducing molecular machines by Jean-Pierre Sauvage, Sir. J. Fraser Stoddart and Bernard L. Feringa. Because of its interdisciplinary nature, Supramolecular Chemistry is now prevalent in many scientific areas such as chemistry, physics, materials and biological sciences.

Published: Eliva Press SRL
Address: MD-2060, bd.Cuza-Voda, 1/4, of. 21 Chişinău, Republica
Moldova
Email: info@elivapress.com
Website: www.elivapress.com

ISBN: 978-1-63648-017-6

Content	Page No

Introduction to Supramolecular Chemistry and Optical Sensing Technology

Shampa Chakraborty[a*], Moumi Mandal[b], Sandipan Halder[c], Sadhana Rayalu[a]

[a] Environmental Materials Division, National Environmental Engineering Research Institute, Nagpur, 440020

[b] Department of Chemistry, Indian Institute of Engineering Science and Technology, Shibpur, Howrah, 711103, West Bengal, India

[c] Department of Chemistry, Visvesvaraya National Institute of Technology, Nagpur 440010, India

E-mail: (Shampa Chakraborty) shampa132@gmail.com

Abstract:

Chemical sensing is an important field for the determination of metal ion, anions, and neutral molecules in recent decades. Sensing method is superior over the other available detection methods. The other method includes flame atomic absorption spectrometry-electro thermal atomization (AAS-ETA), atomic absorption spectrometry (AAS), ICP-AES and flame photometry[3]. These methods are not suitable for analysis of multiple samples although it can give good result, the reason behind is all those methods demand expert hands and superior infrastructure. On the other hand, selective chemical sensors are very useful for the monitoring of lot of samples and the methods are convenient and cheap, easily operable and require no pre-treatment. An ion sensor should be selective for a particular host material but has poor sensitivity to others.

Keywods: Chemosensor, Fluorosensor, Basic concept of sensor, Supramolecular chemistry

1. Introduction

Recognition Chemistry is one of the most fashionable and budding areas of chemistry. Supramolecular chemistry studied the area of protein folding, host-guest chemistry, interlocked molecular architectures, and dynamic non-covalent chemistry. Molecular recognition deals with the detection of ionic or neutral guests by specific interactions with the host molecule. The specific interactions involved non-covalent forces with energy ranging from less than 5 kJ/mol up to 300 kJ/mol. When the formation of host-guest complexes changes the optical or electrochemical properties then, these systems can usually be developed into sensors. Nature is the master of molecular recognition. Pedersen in 1967 first introduced the concept of molecular recognition in abiotic chemistry with his pioneering work on crown ethers. Pedersen used 18-crown-6 as cation binders when he uncovered these compounds in the 1960s. Lehn elaborated these macrocycles into macrocyclic "cryptands" or three-stranded crown ethers that can completely envelop a cation, anion, or other molecular species. Cram's approach to enveloping molecules evolved from monocyclic through bicyclic (jaws) structures to the three-dimensional and highly selective spherands.

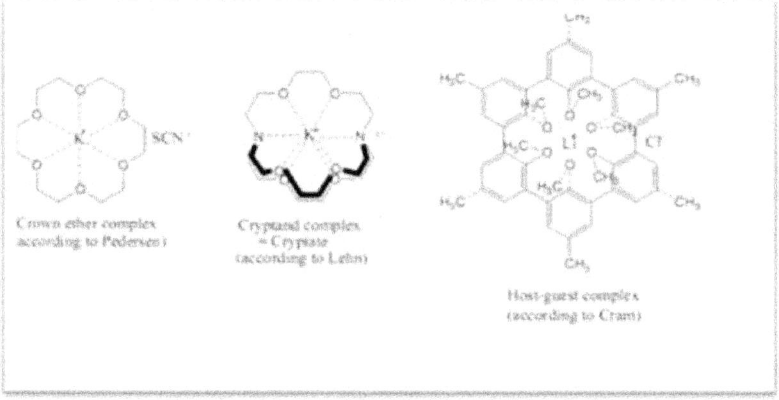

Scheme 1: Structure of different Crown ethers and Cryptands designed by Pedersen, Lehn and Cram

2. Basic Concept of Chemical Sensing

Recognition chemistry has growing interest in the field of supramolecular chemistry because of its vast applications in chemistry, biochemistry, and environmental field. A chemosensor can detect a particular guest species resulting a photophysical change like absorption, emission, or redox potential of the host molecule. In particular a chemosensor composed of two units, receptor unit and another is signaling unit. The receptor unit can selectively bind to the host moieties and the signaling unit can result photophysical perturbation in optical or any other properties.(Figure 1).

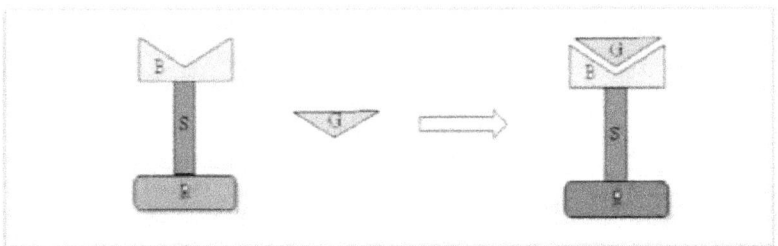

Figure 1. Diagrammatic representation - Here. R= Reporter group: fluorogenic, Chromogenic or Redox active, B= Binding site, S= Spacer- separates binding site and reporter group, G= Guest molecule.

On the other hand, Fluorosensor has two components, an ionophore which is requied for selective binding and a fluorophore which can report the change or perturbation of the host molecule due to the host-guest binding. The change can be fluorescence enhancement or inhibition. The mechanisms which can control the response of a fluorophore to guest binding are photoinduced electron transfer (PET), Fluorescence (Forster) resonance energy transfer (FRET), excimer/exciplex formation or extinction and photoinduced charge transfer (PCT).

2.1 History and Definition:

Molecular recognition[1] may be defined as the specific interaction between two or more molecules through noncovalent bonding such as hydrogen bonding, metal coordination, hydrophobic forces, van der Waals forces, pi-pi interactions, and/or electrostatic effects.[1] According to Emil Fischer's "lock and key" mechanism, the host and guest involved in molecular recognition must be complementary to each other.[2]

Molecular recognition is an important fundamental natural phenomenon. C. J. Pedersen (1967) first enlightens the concept of molecular recognition by synthesizing 18-crown-6 and its derivatives. The ability of these compounds to dissolve alkali and alkaline earth metal salts in organic solvents has lead to the development of large no of synthetic ionophores which have found immense applications and utility.

The unifying observation of Pedersen that cyclic polyether 18-crown-6 tightly encapsulated K^+ ion in abiotic system were brilliantly exploited by other workers most notably by D. J. Cram and J. M. Lehn. They shared the Nobel Prize in chemistry in 1987 and 'Host-Guest' chemistry was formally born with this finding that crown ether serve as a molecular host to the cationic guest.

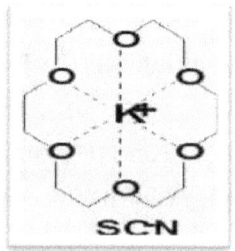 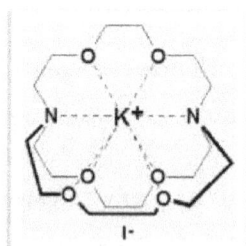 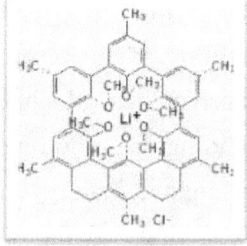

Pedersen's Model **Lehn's Model** **Cram's Model**

Supramolecular chemistry[3] is an interdisciplinary field and this wide horizon is interesting as well as challenging to the chemists for the expression of their creativity and art. This terminology was first used by Lehn has evolved from various sources including inorganic, organic, physical and biological chemistry. The term 'supramolecular chemistry' is usually used in particular for large multiprotein architectures and organized molecular assemblies. Supramolecular chemistry consists of two concerning areas (1) supermolecules, well-defined discrete oligomolecular species; (2) supramolecular assemblies, polymolecular entities, which is made by the spontaneous association of a large undefined number of components into a specific phase.[4] It has more or less microscopic organization as well as macroscopic characteristics e.g. membranes, vesicles, solid state structures etc. Nature provides archetypal examples of the structures, properties and functions, which can be achieved by supramolecular systems.[5]

The concept of molecular recognition is evolving as a rapidly developing science that recognizes weak intermolecular forces and direct molecules towards self-linking as well as hetero linking in supramolecular architecture throughout the different aspects of chemistry of life and material science.

2.2 Molecular recognition in Nature:

In molecular recognition, complementarity is one of the most important concepts which is evident from many biological events e.g. in the DNA-double helix, binding of guanine by ribonuclease T1 etc.

These types of nucleotide binding proteins use both hydrogen bonding and π-stacking force to provide strong and selective complexation. Ribonuclease T_1

binds guanine via two hydrogen bonds between the peptide backbone and O(6) and H(1) groups of guanine.[6]

The stacking interaction (3.4 Å) between a tyrosine residue (Tyr-45) and purine plane (Figure 1.1) increases the binding force. An 'induced fit recognition' is further exemplified from the fact that Tyr-45 swings from the

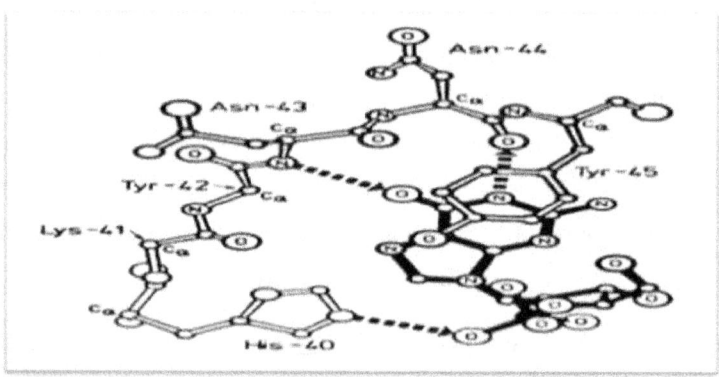

unbound position into a stacking position on guanine binding.[7]

Figure 2.1: X-ray structure of ribonuclease-T1 active site.

The hemoglobin is oxygen carrier in the living system. This dioxygen is used in carbohydrate metabolism and the energy released by this process is further used for the synthesis of ATP. Each hemoglobine unit contains four myoglobine units. Myoglobin is a single-chain protein of 153 amino acids, containing a iron-containing porphyrin i.e. iron (II) is present at the active site, prosthetic group in the center around which the remaining apoprotein folds (**Figure 2.2**). It is the primary oxygen-carrying pigment of muscle tissues[15] whereas hemoglobin is present in blood red corpuscles.[8]

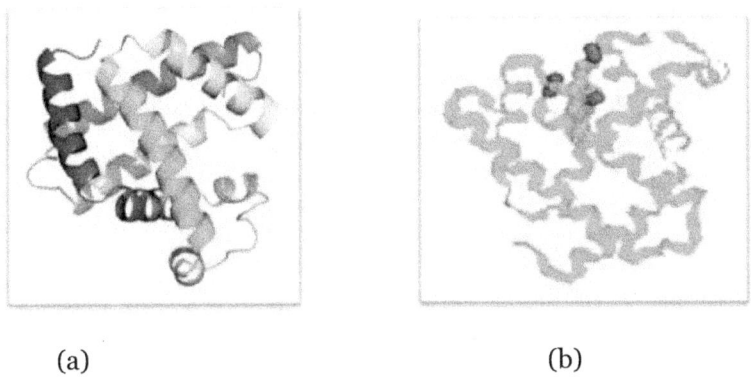

(a)   (b)

Figure 2.2: (a) Model of helical domains in myoglobin; (b) Structure of the myoglobin protein with the position of the heme group highlighted.

2.3   Designing principle of receptors:

Chemistry of artificial receptor molecules represents generalized co-ordination chemistry which is not limited to transition-metal ions but it extends to all types of substrates like cationic, anionic or neutral species of organic, inorganic and biological etc. Molecular recognition implies the shape and size complememtarity and interactional complementarity between the partners which are involved in binding mechanism. Again the complexation strength is dependent upon the number of degrees of freedom occurred during a binding event.[10]

In order to achieve effective binding the following points in designing the receptor are to be taken into account for synthesizing an ideal receptor.

- ❖ Shape and size complementarity, i.e. steric complementarity between the host and guest molecule i.e. presence of convex and concave domains in the correct location,

- ❖ Presence of complementary binding sites which is also called interaction complementarity, such as positive-negative, charge-dipole, dipole-dipole, hydrogen bonding donor-acceptor.

- ❖ Large contact area between receptor-substrate.

- ❖ Multiple interaction sites, which increase the association constant as the number of interaction sites increases.

- ❖ The effect of the medium i.e. the solvent: this factor has marked role because solvent molecules interacts both with guest as well as with the host. Thus hydrogen bond formation between solvent-substrate will be favorable in more polar solvents.

Overall strong binding is required for the high selectivity and high stability of receptor-substrate complex.

Designed hypothetical receptors may be synthesised either by stepwise synthesis of total structure or regio- and stereo-specific functionalisation of certain readily available host molecules. To simplify the long experimental work, it is convenient to make specific molecular design having limited number of recognition elements. Structural units have been incorporated that may respond to or be perturbed by external factors. Most investigations are focused on endo-receptors in which the binding sites are oriented into a molecular cavity; exo-receptors with outward directed sites[11] Steric complementarity between host-guest is also necessary for binding to be favourable. Steric clashes reduce the binding energy, and, especially in chiral systems any form of effective binding may be prevented. This can be advantageous for example enantiomer discrimination is possible.

Molecular polarizabilities, dipoles and conformations dictate the properties of solvents, the interaction of solutes with the solvent and the interactions between solutes. Since the vast majority of the reactions preferred by organic chemists occur in solution, the choice of solvent can play an extremely important role in controlling the reaction. If the solvent strongly solvates either host, guest or complex, or if the solvent interacts strongly with itself, then this can have dramatic effect on the host-guest equilibrium.

Typically, interactions are considered in two groups: ionic/dipolar which are saturated by polar solvents. These effects therefore allow emphasis to be placed on either class of interactions experimentally, provided adequate solubility can be achieved. For example, the conformation adopted by crown ethers is strongly dependent on the nature of the solvent (Figure 2.3).[12]

Figure 2.3: The preferred conformation of crown ether is strongly solvent-dependent. 18-Crown-6 can solubilise potassium ion in the organic phase (left) and is, itself, water-soluble (right).

No matter how many favourable interactions can be arranged between host-guest, binding will not occur to an appreciable extent if the energy released by them fails to overcome the loss of entropy incurred.[13] When this is effected in an appropriate manner, the optimal binding conformations of host and guest lie within the reduced conformational space of the constrained system (Figure 2.4).

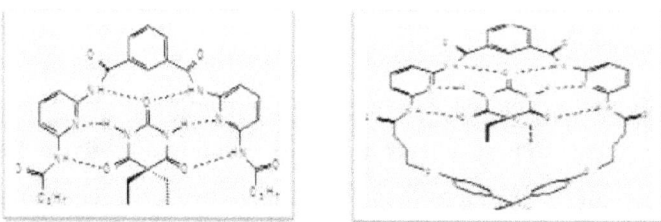

Figure 2.4: The constrained receptor (right) binds barbital (shown) more strongly than the acyclic analogue. Less entropy is lost in the preorganised system when the six hydrogen bonds are formed.

2.4. Designing principle of guest:

Guest molecules are taken into consideration for certain host molecules keeping all the possible recognition elements in mind for the 'best fit' interaction which may be achieved by doing some theoretical computational method. However, based on host molecules, satisfactory screening of guest molecules is possible either via force field computation or via empirical investigation using a CPK molecular model.[14]

2.5 Noncovalent forces in Host-Guest interactions:

Supramolecular chemistry forms the basis of non-covalent interaction. It involves an enormous range of attractive and repulsive forces. The following interactions should be taken into account:

(a) Electrostatic interaction which may be of different types e.g. ion-ion interaction (100-350 KJ/mole), ion-dipole interaction (50-200 KJ/mole) and dipole-dipole (5-50 KJ/mole) interaction.

(b) Hydrogen bonding (4-120 KJ/mole) interactions

(c) Cation π (5-50 KJ/mole) and π-π stacking (0-50 KJ/mole) interactions

(d) Vander Waals forces which is less than 5 KJ/mole

(e) Hydrophobic effect which includes the effect of solvent

(f) Close packing in the solid state

Non-covalent forces are weak forces ranging from 2 KJ/mole for dispersive forces to 20 KJ/mole for a hydrogen bong to 250 KJ/mole for an ion-ion interaction. Hydrogen bonds are omnipresent in supramolecular chemistry. It is also responsible for the overall shape of many proteins' recognition of substrates by numerous enzymes and for the double helix structure of DNA.

Electrostatic interactions involve Coulombic attraction forces between two opposite charges. Ion-ion interactions (**Figure 2.5**) are basically non-directional, but in case of ion-dipole interactions the dipole must be suitably aligned for optimal binding efficiency. Thus the high strengths of electrostatic interactions have been used by supramolecular chemists to achieve strong binding. Based on this electrostatic interaction, many receptors for cations (crown ethers, cryptands and spherands) and anions (protonated or alkylated polyammonium macrobicycles) have been employed to hold the guest in place.

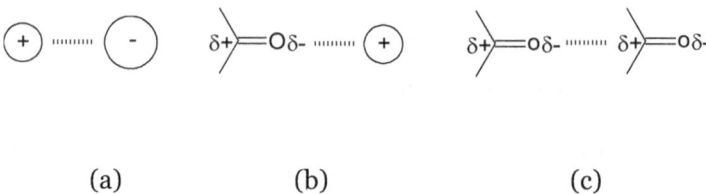

(a)   (b)   (c)

Figure 2.5: Electrostatic interactions (a) ion-ion, (b) ion-dipole, (c) dipole-dipole.

Hydrogen bonding is very important in many biological systems (e.g. the DNA double helix) and this have been utilized in receptors designed to co-ordinate neutral organic species such as barbiturates, short chain alcohols and amides, and also in anions. Individual components can be built into molecular designers with the precision and directional nature of the hydrogen bonds which facilitates the construction of complex architectures.

π-π **stacking forces** are mainly observed in systems containing aromatic rings. Attractive interactions can occur in either a 'face-to-face' or 'edge-to-face' or in 'offset' manner. Current theories have suggested that these attractive forces are electrostatic in nature. Some elegant receptors have been synthesized employing this π-π stacking interactions, including a receptor for benzoquinone.

The **dispersion forces** (or induced dipole-induced dipole interactions) are attractive forces between molecules of instant dipoles or in the electron clouds around and each molecule, interact suitably. The van der Waals forces provide additional enthalpic stabilisation to the coordination of a hydrophobic guest into the hydrophobic cavity. Because of its very general nature it is difficult to design receptors specifically to take full advantage of them. One such system may be a self-assembled 'tennis-ball' that can encapsulate xenon atoms.

Hydrophobic effect is the specific driving force for the association of non-polar binding partners in aqueous solution. Water molecules around the non-polar surfaces of a hydrophobic cavity arrange to form a structured platform. Upon guest complexation, the water molecules are released and become disordered which results in increase in entropy. Again the hydrogen bonds between water molecules are larger than the interactions between the water molecules and non-polar solutes, providing an enthalpic force for apolar guest coordination (i.e. when the water in the apolar cavity is released into the bulk solvent it can maximize its hydrogen bonding interactions). Receptors containing hydrophobic interior cavities designed to encapsulate organic guest molecules in aqueous solution include the cyclophanes and cyclodextrins. Thus this is an enthalpically as well as entropically favorable process.

Donor-acceptor ability also plays an important role in hydrogen bonding recognition. A good electron pair donor will readily accept a hydrogen bond. If the solvent is a good hydrogen bond donor or acceptor, this disrupts and weakens the recognition process, which depends on hydrogen bonds, because the unbound host and guest are solvated more effectively than that in their bound form. This problem is particularly more prominent in aqueous solution (as water is both an excellent hydrogen bond donor and acceptor), water molecules always have a competition with the guest substrates to be bound with the host molecules.

Close packing in the solid state is also a driving force in the solid state crystalline structure. According to the theory of close packing introduced by Kitaigorodsky,[15] it is simply a manifestation of the maximization of favorable isotropic vander Waals interactions.

Organic molecules exhibit maximum entropy in solution phase as they possess conformational flexion and chaotic motion. Opposing this behavior is the formation of non-covalent bonds, which reduces the system's overall free energy. This relationship is described by Gibbs' law, $\Delta G = \Delta H - T\Delta S$.

2.5.1. Hydrogen bonding interaction:

By virtue of its directionality, specificity and biological relevance, hydrogen bonding interaction is one of the favorite intermolecular forces in molecular recognition studies. This concept of hydrogen bonding was first formulated in about ninety years ago and has found immense application in explaining biological events.

Different arrangement of donor-acceptor array has been studied so far leading to formation strong host-guest complexes.[16] Besides this base-pairing of nucleic acid is primarily governed by the hydrogen bonding to carry the unique genetic code.[17]

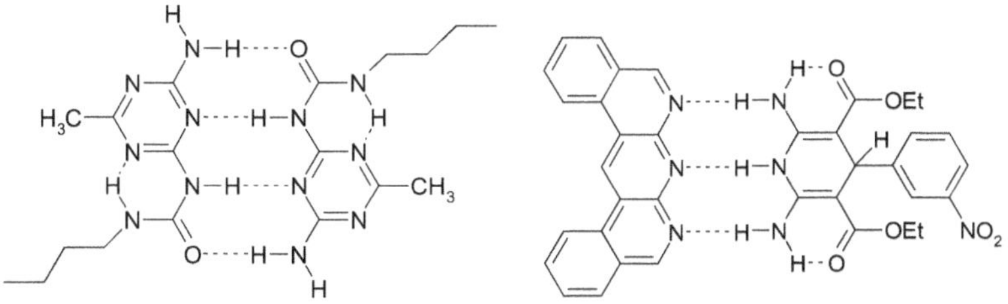

Figure 2.6: Hydrogen bonding array in synthetic donor-acceptor system: (a) DADA-ADAD mode of homo inteaction; (b) AAA-DDD mode of hetero interaction.

2.5.2. π-stacking interaction:[18]

Most binding forces involve simple electrostatic attractions at their origin. Therefore regions of negative charge will in general be attracted to the regions of positive charge. π-stacking interaction is one of the most used terms in molecular recognition. Simple aromatics experience favourable interactions with each other. Simultaneous participation of both hydrogen bonding and π-stacking interactions offer a rationally powerful approach for the effective stabilization of the molecular complex. This π-stacking force has been used recently for the recognition of many aromatic acids,[19] amino acids,[20] and planar heterocycles such as adenine, thymine etc.[21]

The geometry of the π-stacking interacting groups is important in DNA and protein crystal structure.[23] The T-shaped or edge-to-face geometry is better for some cases. Where T-shaped geometry is not obtained, the best form will be the displaced or slipped stack. These types of models have been proven to be very effective in explaining many binding phenomenon taking place between the host-guest in molecular recognition process.

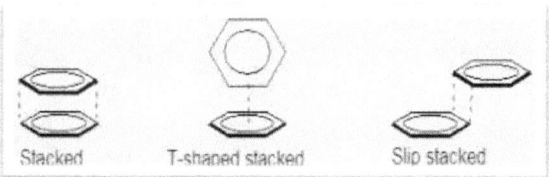

Although the exact nature and origin of π-stacking effect (either charge transfer or simple Vander Waals forces) is still not clear, electron donor model which is based on experimental findings suggests that the strong attraction is raised because of electronic interaction between an electron donor and an electron acceptor. According to the charge transfer model, a CT complex is formed from the association of good electron donor and electron acceptor and this is characterized by CT transition band in the UV-vis absorption spectrum.

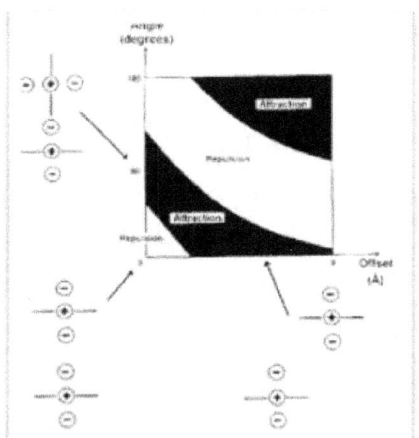

Figure 2.7: Interaction between two idealized π-atom as a function of orientation: two attractive geometries and the repulsive face-to-face geometry are illustrated.

Yang[22] et. al. have shown π-stacking distyrylbenzenes (**Figure 2.8**) resulting in highly ordered assembly.

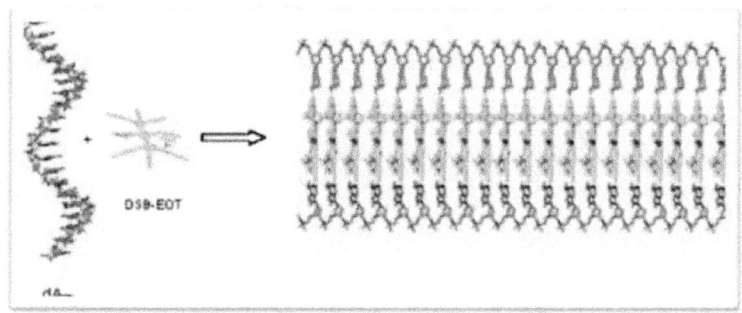

Figure 2.8: π-stacked assembly of distyrylbenzenes.

2.6 Development and applications:

2.6.1. Liquid crystals:[23]

Liquid crystals are substances which have properties between those of a conventional liquid, and those of a solid crystal. For instance, a liquid crystal (LC) may flow like a liquid, but have the molecules in the liquid arranged and/or oriented in a crystal-like way. Here also π-stacking interaction plays an important role for the formation of the liquid crystals.

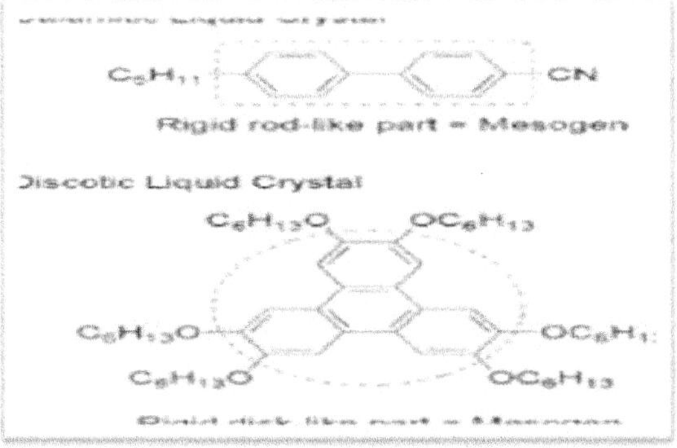

Figure 2.9: Formation of liquid crystal through π-π interactions.

Liquid crystals have wide application and use in liquid crystal displays, which rely on the optical properties of certain liquid crystalline molecules in the presence or absence of an electric field. Liquid crystal in fluid form is used to detect electrically generated hot spots for failure analysis in the semiconductor industry. Liquid crystal memory units with extensive capacity were used in space shuttle navigation equipment.

2.6.2. Drug design:

The ability of a particular molecule to recognize selectively even in presence of closely related partners is the basic principle of many processes involving chemical reactivity, enzyme catalysis, gene regulation etc. The binding of small molecules to complex proteins forms the basis of modern drug design. Drug molecules are specifically designed to recognize only those particular enzymes or receptors involved in the disease process. Such drugs must be highly discriminating at the molecular level so that disease can be controlled without side effects. Many drugs in racemic form are available for clinical use. Different stereoisomers of drugs can cause different physiological responses, e.g. d-propanolol is approximately 40 times more efficient than its l form and acts as antiarrhythmic and antihypertensive agents whereas l-propanolol is an effective anti angina pectoris[24] agent. β-cyclodextrin can effectively bind d-propanolol from the racemic mixture of d and l isomer through hydrogen bonding and thereby separation of these two enantiomers (chiral recognition)[25] is possible.

Drug design is an iterative process. It begins when a chemist identifies a compound that displays an interesting biological profile and ends when both the activity profile and the chemical synthesis of the new chemical entity are optimized. Traditional approaches to drug discovery rely on a step-wise synthesis and screening program for large numbers of compounds to optimize activity profiles (**Figure 1.10**). Over the past ten to twenty years, scientists have used computer models of new chemical entities to help define activity profiles, geometries and reactivity.

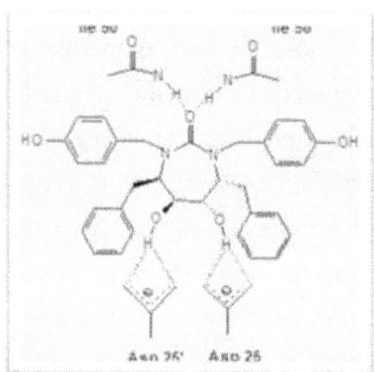

Figure 2.10: Inhibitor for HIV protease bound in the enzyme active site.

2.6.3. Binding of anions:

The field of non-covalent anion coordination chemistry as we know it today may be traced back to a report by C. H. Park and H. E. Simmonds in 1968, concerning the halide complexation properties of a series of macrobicyclic hosts termed katapinaands (Greek katapino, meaning to swallow up or engulf). The selective binding of anionic guest species is challenging than binding metal ions, even though qualitatively the same kinds of concepts and ideas that span the whole of supramolecular host-guest chemistry should apply.

Many research groups have worked in the field of anion recognition, a few examples are colorimetric receptor for fluoride anion by Bozdemir and co-workers[26] and colorimetric as well fluorescent receptors for BO_3^- by Choi et. al.[27]

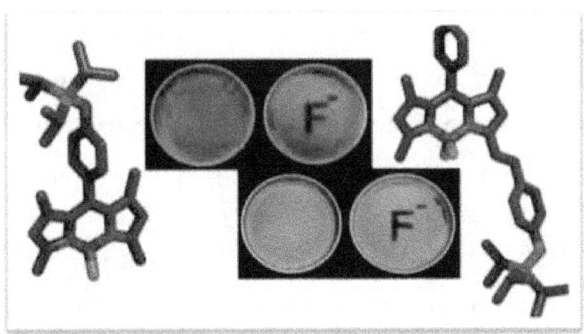

Figure 2.11: Colorimetric detection of fluoride anion by Bozdemir et. al.

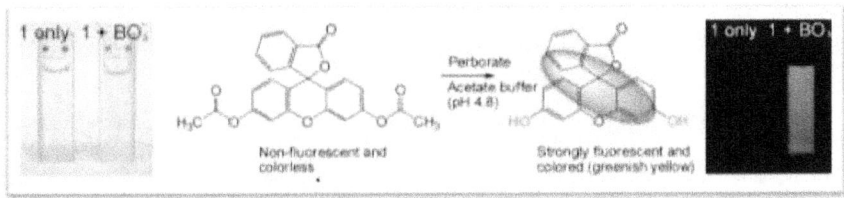

Figure 2.12: Recognition of BO_3^- by Choi et. al.

2.6.4. Binding of cations of significant environmental concern:

The thermodynamic selectivity of a given host for a particular guest cation represents the ratio between the affinity of host and other guest cation. Thus a successful host exhibits a strong affinity for one particular guest and a much lower affinity for other cations. Thus designing a synthetic host that will be highly selective for a given cation is a very complicated task because the selectivity is governed by an enormous number of factors involving size complementarity between host and guest cation, electrostatic charge, solvent polarity, degree of host preorganization etc. Selected examples regarding the recognition of cations may given as sensing of Cu^{2+} through luminescence by Chakraborty et. al. [28] (Figure 1.13) colorimetric, fluorescence and electrochemical receptor for Pb^{2+} by Zapata et. al.,[29] (Figure 1.14) colorimetric receptor for Cu^{2+} by Sheng et. al.,[30] (Figure 1.15) etc.

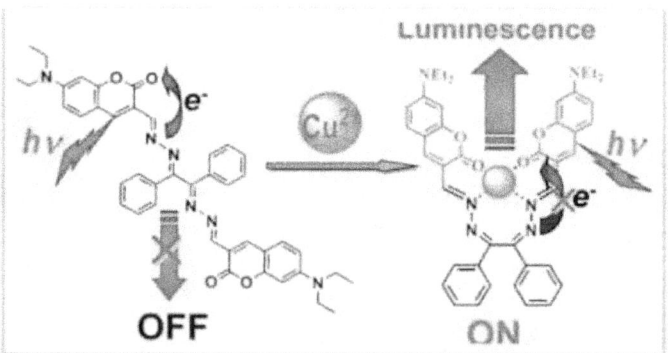

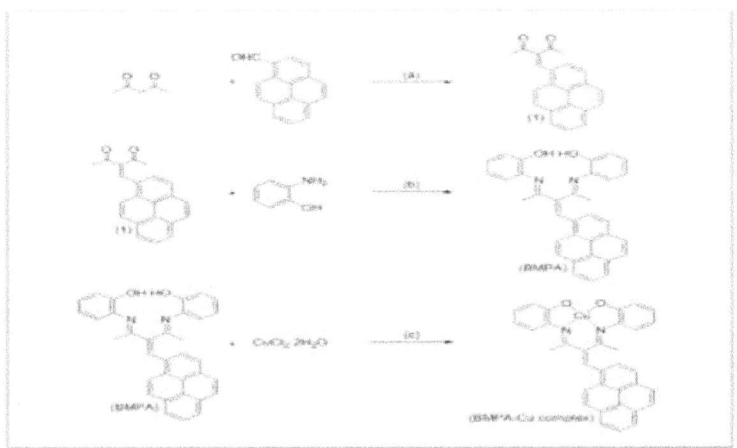

Figure 2.13: Fluorescence recognition of Cu^{2+} by He et. al.[28] and Chakraborty et.al[28]

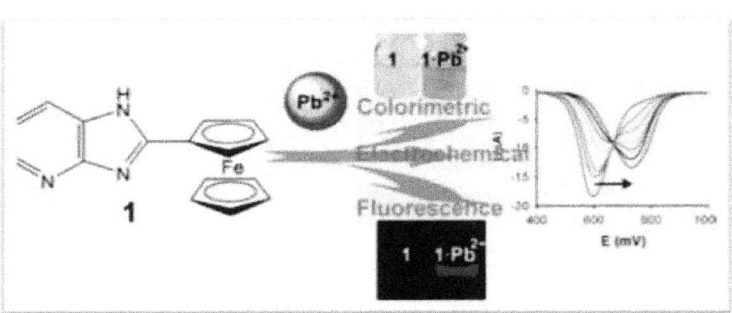

Figure 2.14: Colorimetric, fluorescence and electrochemical receptor for Pb^{2+}.

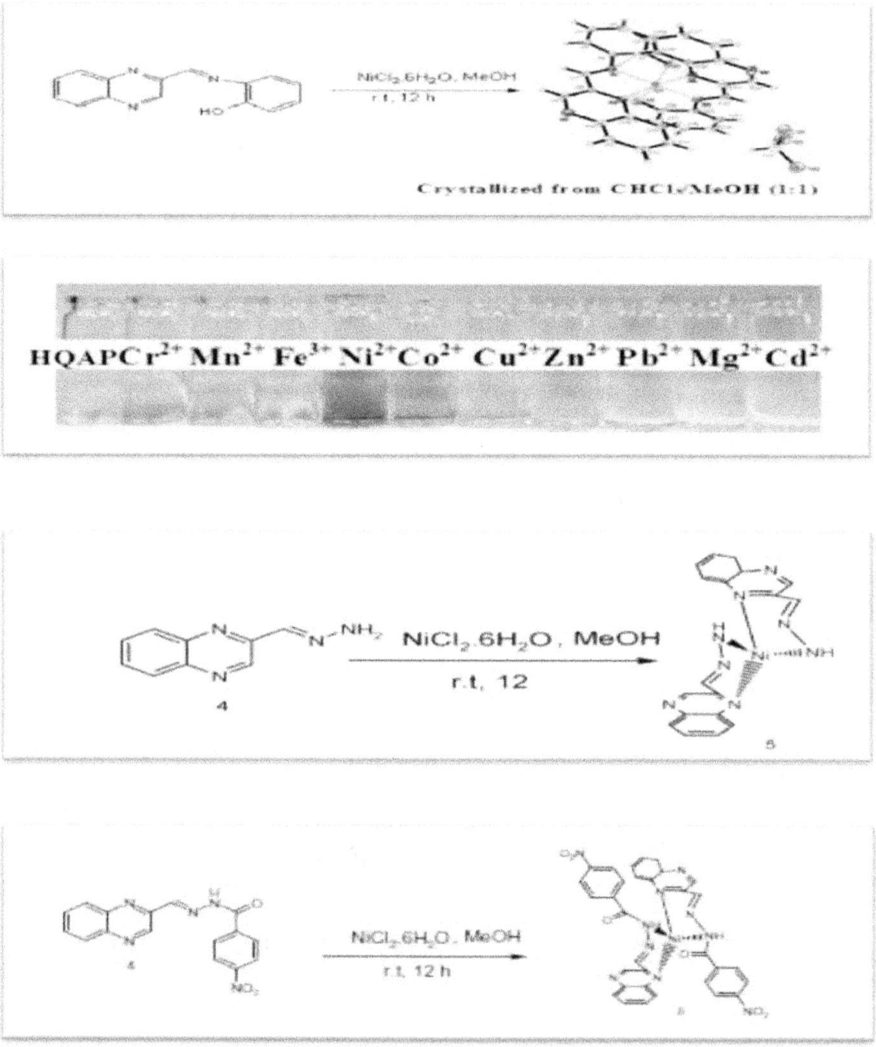

HQAPCr^{2+} Mn^{2+} Fe^{3+} Ni^{2+}Co^{2+} Cu^{2+}Zn^{2+} Pb^{2+} Mg^{2+}Cd^{2+}

Figure 2.15: colorimetric receptor for Ni^{2+}.[29]

2.6.5. Molecular sensors:

Fluorescence sensing of chemical and biochemical analytes is an active area of research.[31] This research is being driven by the desire to eliminate radioactive tracers, which are costly to use and to dispose of. There is a need for rapid and low-cost testing method for a wide range of chemical, bioprocess and environmental applications.

Figure 2.16: Schematic representation of molecular sensor.

Sensors should ideally be selective for a particular guest and not only report the presence of the guest molecule, but should also allow the chemist to monitor its concentration. This quantitative analysis by fluorescence sensor is medically and environmentally important. An example of artificial sensor for Hg^{2+} has been given by Zhou et. al.[32] (Figure 1.17) where sensing of mercury ion has been achieved by strong excimer formation after binding with Hg^{2+}.

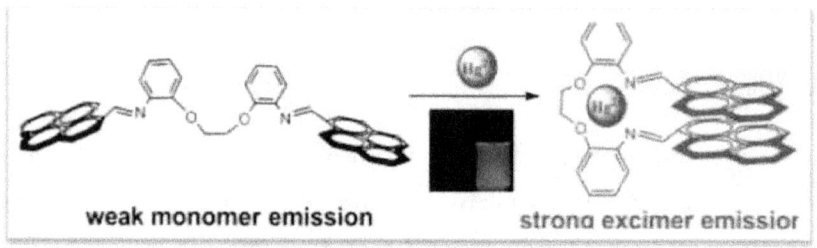

weak monomer emission   strong excimer emission

Figure 2.17: Sensing for Hg^{2+}.

2.6.6. Switches and molecular machinery:

Switches and molecular machinery technology use individual molecules for controlling and storing information, acting as 'on-off switches' and 'logic gates'. If the molecule contain an electron donor group like amine, then the phenomenon will be more exciting. This phenomenon has been exploited to develop sensors bound on quenching of polynuclear aromatic hydrocarbons by amines.[33]

The basic idea is that quenching by amines requires that the lone pair is bound to a proton, then electron transfer is inhibited and the fluorescence is not quenched. Such probes are said to undergo photoinduced electron transfer from nitrogen into the aromatic ring. Preventing PET, in similar way the emission from the anthracene unit is observed. So protons switch on the fluorescence as shown in Figure 2.17.

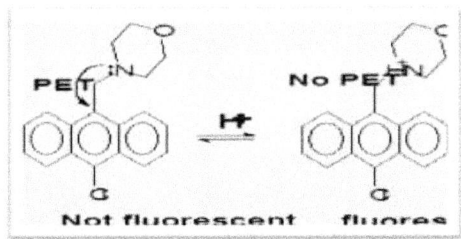

Figure 2.18: Molecular 'on-off' switch.

2.6.7. In catalysis:

Bisubstrate reaction template:

Kelly et al. have made a different approach for acheiveing catalytic effect in receptor-mediated bond forming reaction.[34] The receptor molecule has two sites to bind two substrates, which are capable of reacting with each other as shown in Figure 1.19. The appropriate steric disposition of the two reacting functionalities and of the intramolecularity of reaction gives the product easily with six-fold rate acceleration. This is obviously based on hydrogen bond mediated stabilization of the transition state.

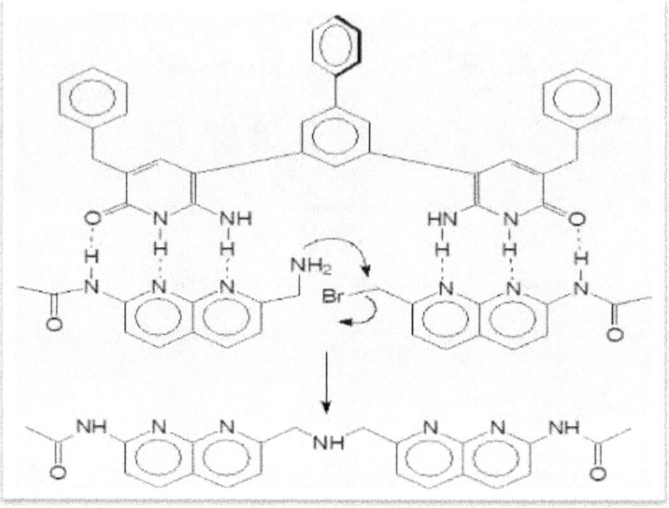

Figure 2.19: Hydrolysis of adenosine triphosphate:

It was found that protonated macrocyclic polyamines can catalyse the hydrolysis of adenosine triphosphate (ATP) to adenosine diphosphate (ADP) (Figure 2.20). Cyclic hexaammonium action strongly binds ATP and an intramolecular transfer of the terminal phosphate to the sixth nitrogen gives an intermediate, which then loses phosphate ion by reaction with water.[35]

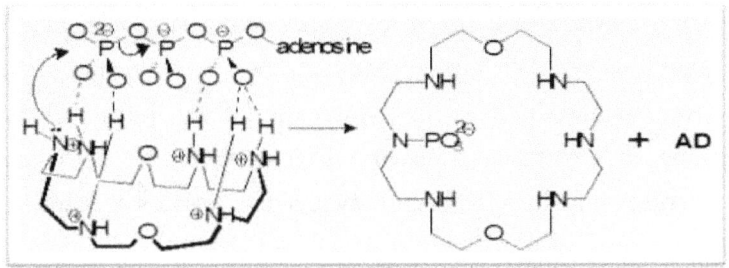

Figure 2.20: Hydrolysis of adenosine triphosphate.

2.6.8. In transport process:

Now-a-days the development of low molecular weight anion transporters has got the attention of many scientists. The special distribution of most abundant anions viz. chloride, bicarbonate and phosphate, in physiological system, is not even. For example the intracellular and extra cellular concentrations of chloride ion are typically 5-15mM and 110mM respectively. If we consider the case of unassisted ion transport, we can see that the membrane permeability co-efficient for monovalent halide anions is 2-3 orders of magnitude faster than the corresponding cation of similar size, e.g. K+. Although the reason is not clear it has been proved that the cations are absorbed more strongly into the phospholipids bi-layer negative ends compared to that of the anions. These negative ends restrict the halide anions also due to repulsion. But the presence of net positive charge into the cell anions gets transported into the cell.

The need of the synthetic anion transporters comes from the fact that due to the inabilities of the family members of the corresponding anion transporters, various diseases like pancreatic failure, intestinal blockage, cytic fibrosis, kidney stones, muscle stiffness etc. occur and also a few number of compounds has the capability to act as carriers of anions.

This goal can be achieved by two ways

(I) By using anion transporters

(II) By developing anion channel.

Sessler et. al. used dipyrrole and tri-pyrrole based compounds for this purpose and Gale et. al. used pyrrole carboxamide moiety for synthetic anion carriers. In these cases after protonation of the designed carriers they become very good carriers of Cl⁻ ion.

2.6.9. Recognition through metal-templated self-assembly:

Self assembly is commonly observed in nature e.g. in biological systems, the important aspartate proteinase HIV protease receptors which undergoes a self dimerisation to bring together the key aspartate carboxylate moieties in its active site. A related example is seen in the formation of the antigen recognition site in antibodies.

2.6.10. Nanotechnology and molecular devices:

Nanotechnology[36] is a field of applied science and technology covering a broad range of topics. The main unifying theme is the control of matter on a scale smaller than 1 micrometre, normally approximately 1 to 100 nanometers, as well as the fabrication of devices of this size. It is a highly multidisciplinary field, drawing from fields such as applied physics, materials science, colloidal science, device physics, supramolecular chemistry, and even mechanical and electrical engineering. Examples of nanotechnology in modern use are the manufacture of polymers based on molecular structure, and the design of computer chip layouts based on surface science. Despite the great promise of numerous nanotechnologies such as quantum dots and nanotubes, real commercial applications have mainly used the advantages of colloidal nanoparticles in bulk form, such as suntan lotion, cosmetics, protective coatings, and stain resistant clothing.

2.6.11. Pharmaceuticals:

The concept of supramolecular chemistry has been applied to design drugs which can prevent some important disease in the field of pharmaceutical research.

2.6.11.1. Anti-cancer agents:

Texaphyrin is an expanded porphyrin synthesized by Sessler et al.[37] The gadolinium complex Gd-Tex is currently undergoing trials for use as a radiation sensitizer (**Figure 1.21**). A cancer patient is exposed to radiation. When Gd-Tex is exposed to ionising radiation in vivo, it captures an electron and becomes a π-radical cation.

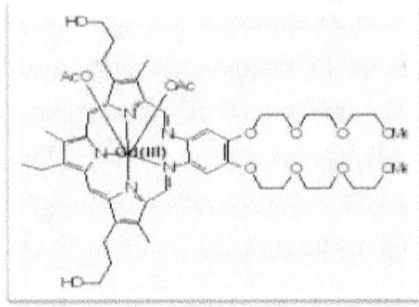

Figure 2..21: Texaphyrin metal complexes of Gd-Tex.

2.6.11.2. **Anti-HIV agents:**

Bicyclam, a receptor for two transition metal cations, exhibits potent inhibition of the HIV virous at early stage in its replication cycle. It is possible that this inhibition is mediated by transition metals.

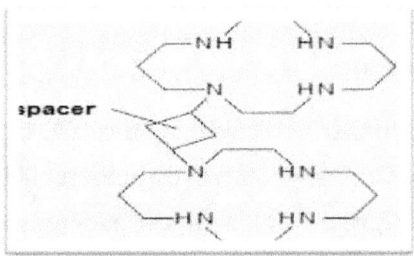

Figure 2..22: Linked bicyclam.

2.6.12. Physico-chemical methods in the study of non-covelent complexation:

The different methods in the characterization of host-guest complex are briefly described below.

NMR is a useful technique to ascertain the host-guest interaction. NMR methods offer a lot of advantages. These are nondestructive analytical methods, and are easy to handle and well suited to study dynamic processes. In this method, a stock solution of known concentration of host and guest in $CDCl_3$ were prepared by accurately weighing pure substrates. Generally the guest concentration was kept much higher compared to host concentration and added to the host solution with varying amounts. The change in chemical shift values of the key protons involved in hydrogen bond is noticed and used to ascertain the binding interaction between the host and the guest. When the broad range of binding constants is considered and dynamic exchange between free host and complex is fast on the NMR time scale, Job method is usually adopted for the stoichiometry of the complex. The stoichiometry of the complexes can also be determined from the break of the titration curve. For the large molecular complexes, it is very difficult to characterize by NMR spectroscopy, as most of the proton spectrum is uninterpretable. These NOESY NMR provides information about the connectivity of the host-guest complexes in the solution.

In addition, mass spectroscopy technique also infers the direct detection of the various species formed in solution, but the ionization method must be mild. Otherwise the complex will be broken into its constituent species, rather than flying through the spectrometer as a discrete unit.

The most convincing evidence of a supramolecular interaction is a crystal structure of the host-guest complex. Crystallography clearly shows the binding site and the complex is as planned by the molecular designer. A crystal structure is only valid for the solid state, as factors such as crystal packing may alter the properties of super molecule.

In a similar way, UV-vis and fluorescence spectra are especially effective for investigations of π-electron system or transition metals or their spectra, which can be strongly perturbed on complex formation. For both methods it is a straightforward matter to relate the concentration of a species to the intensity of the signal associated with it. An advantage of these approaches is that they can be quite sensitive and disadvantageous and that there are only limited guest with desirable spectroscopic properties, which are amenable to study.

Vapour pressure osmometry, membrane osmometry and gel permeation chromatography provide useful informations about the molecular weights with degree of errors. They only provide a brief account of an average of all species in solution.

3. Classification and design of chemosensors

A chemosensor is a Donor-Acceptor system which contained of electron-donating (ED) and electron accepting groups (EW) at appropriate positions. The particular metal ion will bind ED or EW group depends upon the SHAB concept of the hardness and softness of the binding sites and the guest. When a metal ion binded with ED group the donating ability is decreased, which converted the D-Π-A system to A-Π-A system reducing the conjugation and blueshift has occurred in the absorption spectrum.(**Figure 2).**

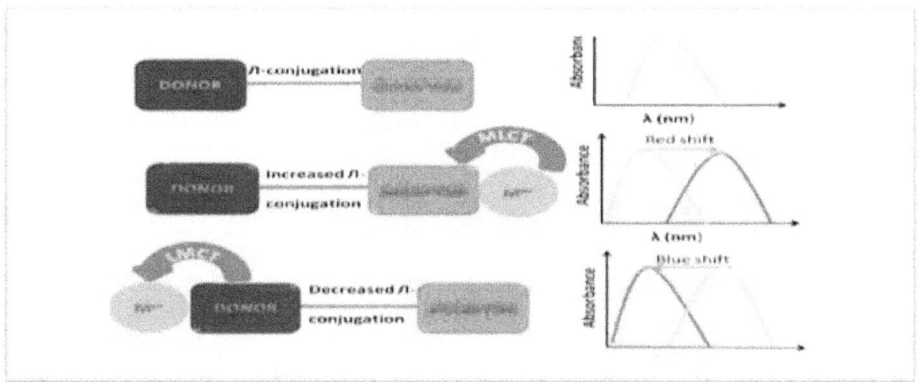

Figure 2: Pictorial presentation of the effect of binding of cation on D- Π -A system and on absorption spectrum.

There are three recognition techniques are used for detection of molecules:

- **binding site-signalling process**
- **displacement technique**
- **chemodosimetric technique**

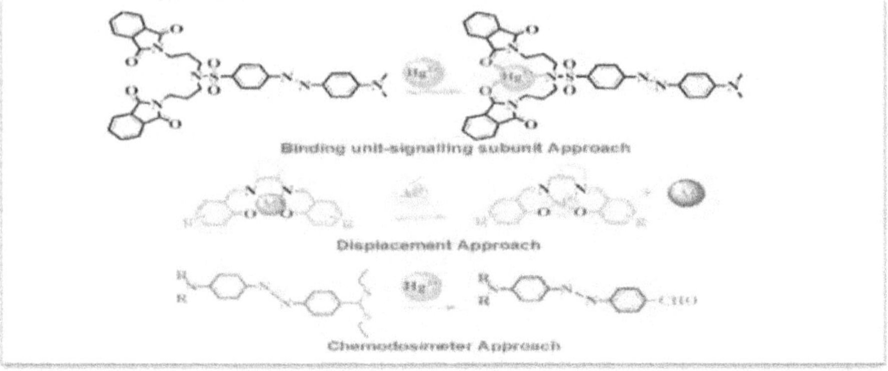

Figure 3: Different techniques employed to achieve colorimetric sensing of metal ions.

The first mechanism is most popular among all the methods for colorimetric change (Figure 4), whereas the displacement technique has been seen by anions by replacement of a bound metal ion. Chemodosimeters has limitations due to the irreversible working mechanism. In the case of displacement technique, some metal-complexes of Schiff base ligands have shown selectivity towards specific metal ions.(Figure 3)

4. Mechanisms involved in Chemical sensing

There are three mechanisms involved for colorimetric probes via different intra and intermolecular Charge Transfer (ICT) which are as follows

- ❖ Ligand to metal charge transfer LMCT
- ❖ Metal to ligand charge transfer (MLCT)
- ❖ Twisted internal charge transfer (TICT)

On the other hand,the mechanisms responsible for fluorometric sensing are:

❖ **Charge transfer (CT) mechanism**

❖ **Electron transfer (ET) mechanism**

❖ **Energy transfer (ET) mechanism**

❖ **Excited-state intramolecular proton-transfer (ESIPT) mechanism**

❖ **Aggregation-induced emission (AIE) mechanism**

❖ **C=N isomerization mechanism**

Charge transfer (CT) Mechanism

Intramolecular charge transfer (ICT) are generally two types, metal–ligand charge transfer (MLCT), and twisted intramolecular charge transfer (TICT). For an ICT-based chemosensor, enhancement of ICT can leads to red shift whereas suppression can lead to a blue shift in its emission spectrum. MLCT, is observed in transition metal complexes whereas charge can transfer a transition metal to a ligand e.g. ruthenium, rhenium and iridium, etc. In addition, TICT is a strong intramolecular charge transfer which can takes place in excited state where solvent relaxation takes place resulting a rotation of electron donor and acceptor until upto it twisted to 90^0.

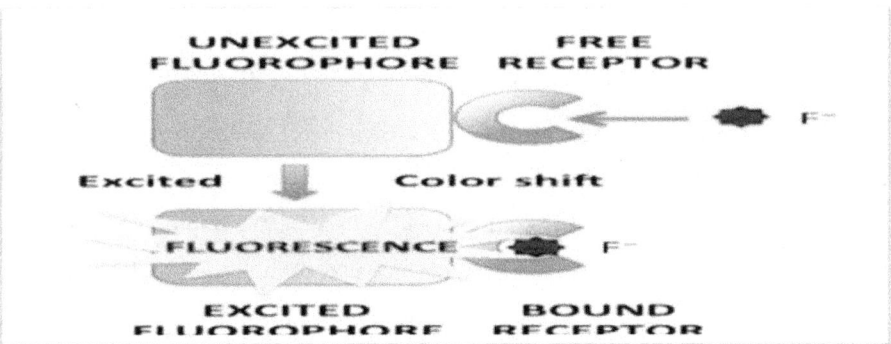

Figure 4. Schematic diagram for ICT mechanisms

Here, it is the design of fluorescent fluoride sensor, possessing an electron-donating group connected to an electron-withdrawing group without spacer. (Figure 4)

Electron transfer (ET) Mechanism

ET is mainly photo-induced electron transfer (PET) which is mainly applied in fluorescent sensors. Generally, the fluorescence is quenched due to the PET process, and recovered by the inhibition of this process by guest molecules.(Figure 5)

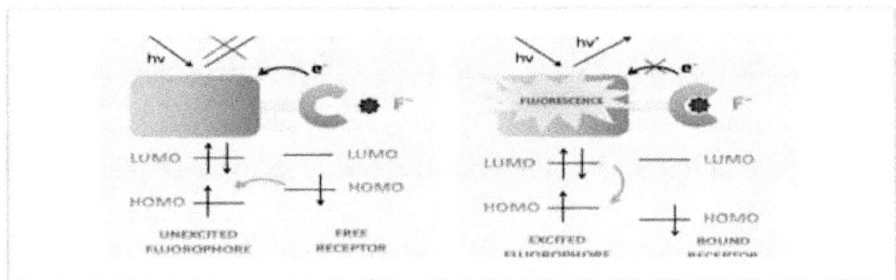

Figure 5. Schematic diagram for PET mechanisms

A PET fluorescent fluoride probes has shown here. A relatively high-energy electron pair without bonding is required in the receptors. After excitation, an electron of HOMO is promoted to LUMO, in which a rapid intramolecular electron transfer occurs between the HOMO of receptor and the LUMO of excited fluorophore in the unbound state. The quenching effect of the probe is in the fluorescence "off" state. However, the bounded receptor caused electron pair coordinating to F- , making the HOMO of receptor even become lower than that of the fluorophore. The perturbed receptor redox potential slowed down or even switched off the PET process, then causing fluorescence"turned on".

Energy transfer (ET) Mechanism

Energy transfer mechanisms are two types electronic energy transfer (EET) and fluorescence resonance energy transfer (FRET). The classifications was done based on the interaction distance between the energy donor and energy acceptor. EET or Dexter electron transfer requires 10 A^0 of distance between donor and acceptor. On the other hand, FRET requires a certain degree of spectral overlap between the emission spectrum of the donor and the absorption spectrum of the acceptor. The distance between the donor and acceptor should be from 10 to 100 A^0 for efficient FRET to occur. (Figure 6)

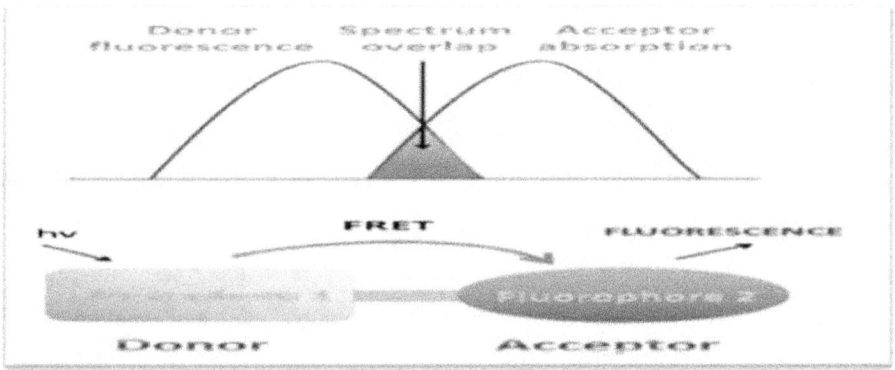

Figure 6. A. Schematic representation of the interaction of two different fluorophores.

Figure 7 showed an example of FRET mechanism. There is a spectral overlap between the emission spectrum of Donor and the absorption spectrum of Acceptor for FRET to occur. FRET happens with an energy transfers from the excited Donor to the Acceptor (D* + A→D + A*), which is coupled in resonance, causing light emission from the acceptor, accompanied with a loss of emission from the donor.

Excimer/Exciplex Formation

If a fluorophore in the excited state has the same structure in the fluorophore in the ground state, then the complex formed is called an excimer. On the other hand, If the fluorophore in the excited state is different from the fluorophore in the ground state, the resulting complex is called an exciplex. Excimer/exciplex formation with a guest results in sensing simply by monitoring the excimer/exciplex band. However, exciplex formation is very rare phenomenon.

Aggregation-induced emission (AIE)

Qunnching of fluorescent emission of organic fluorophores in aggregated form, is denoted aggregation-caused quenching (ACQ). The cause of AIE is the restriction of intramolecular rotation in the aggregates and emission is thus greatly enhanced. Unhindered intramolecular rotation in AIE molecules in the free state leads to efficient nonradiative decay of the corresponding excited states, making them nonemissive. Some nonfluorescent organic molecules in solution were become strongly fluorescent upon aggregation. The factors involved in aggregation of AIE molecules by guest molecules are electrostatic interaction, coordination interaction, hydrophobic interaction, steric hindrance or the influence of polarity and viscosity. Variety of new AIE-active fluorescent bio/chemosensors have been developed to detect ionic species (Hg^{2+}, Ag^+, CN^-), biomolecules (protein, ATP and DNA) and gases and explosives (CO_2, TNT, picric acid).(Figure 7)

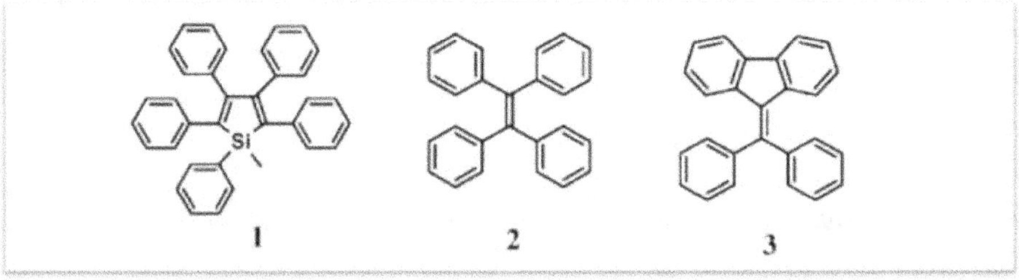

Figure 7: Structures of AIE-active molecules.

C=N isomerization

C=N isomerization is based on the study on photophysical properties of conformationally restricted compounds. It is the decay process of excited states in compounds with an unbridged C=N structure so those compounds are often nonfluorescent. Thus C=N isomerization could be inhibited through complexation of a guest to a designed fluorescent-sensing molecule rather than the covalent bridging of the C=N bond. coumarin-derived imine was designed as a novel fluorescent chemosensor that used C=N isomerization as a signal system.(Figure 8)

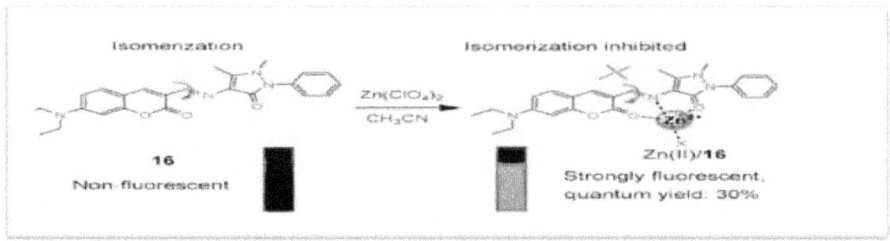

Figure 8: Schematic illustration of the increase in fluorescence achieved by inhibiting C=N isomerization.

Excited-state intramolecular proton-transfer (ESIPT)

ESIPT is a process when an photoexcited molecule relaxed thorugh photoisomerization by transfer of protons in a moleculae if their is an intramolecular hydrogen bonding between a hydrogen bond donor ($-OH$ and NH_2) and a hydrogen bond acceptor ($=N-$ and $C=O$). ESIPT is a unique four-level photochemical process, with the electronic ground state of ESIPT fluorophores typically existing in an enol (E) form. The electronic charge of such molecules can be redistributed upon photoexcitation, and resulted greater acidity for the hydrogen bond donor group and enhanced basicity for the hydrogen bond acceptor within the E form. After that an extremely fast enol to keto phototautomerization takes place. After decaying radiatively back to its electronic ground state, a reverse proton transfer (RPT) takes place to produce the original E form.(Figure 9)

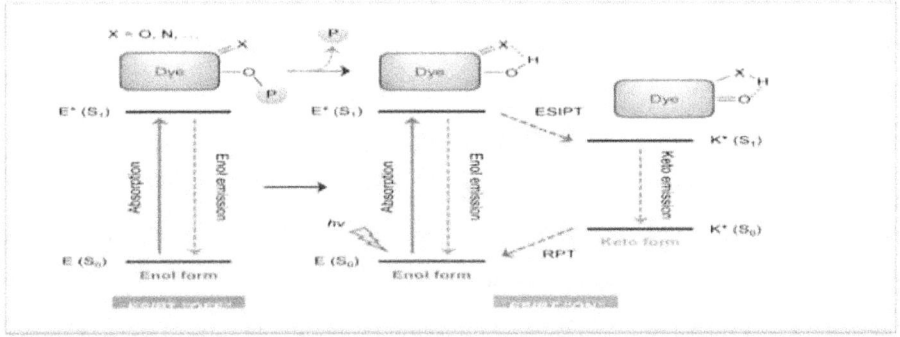

Figure 9: Diagrammatic representation of the ESIPT-based fluorescent probes.

Conclusion

In this book, we have disussed about the chemo and fluorosensors designing principle, different mechanisms involved and the importance of chemical sensor as detection method over other available methods. Molecular recognition is comparatively a new field but it accuired a dominating area and has vast applications in various field of research such as environmental, chemical, biological field. The paper can serve as a good lesson for beginner's of chemical, biological, material and environmental undergraduate and graduate student and researcher.

Acknowledgement

Authors thank the SERB (PDF/2016/002856) (Govt. of India) for financial support. SC acknowledges Director-NEERI for for giving opportunity to work at NEERI-CSIR in National Post-Doctoral Fellowship Scheme.

References:

1.   (a) Hunter, C. A. Angew. Chem. Int. Ed. Engl.**2005**, 43, 5310; (b) Paulini, R. Muller, K.; Diederich, F. **2005**, *44*, 1788.

2.   (a) Lehn, J. M. Supramolecular Chemistry - Concepts and Perspectives, New York, VHC, **1995**; (b) Gellman, S. H. Chem. Rev. **1997**, *97,* 1231.

3.   (a) Steed, J. W.; Atood, J. L. Supramolecular Chemistry, Wiely, **2001**; (b) Davis, J. T. Angew. Chem. Int. Ed. Engl. **2005**, *43*, 668.

4.   Shinkai, S.; Ikeda, M.; Sugasaki, A.; Takeuchi, M. Acc. Chem. Res. **2001**, *34*, 494.

5.   Stryer, L. Biochemistry; W, H. Freeman and Co.: New York, 4th ed.; **1995**.

6.   Heinemann, U.; Saenger, W. Nature, **1982**, *299,* 27.

7.   Fersht, A. in Enzyme Structure and Function, Freeman, Reading, **1977**, 262.

8.   Lodish, H.; Berk, A.; Zipursky, L. S.; Matsudaira, P.; Baltimore, D.; Darnell, J. Molecular Cell Biology, 4th Ed., W. H. Freeman, **2000**.

9.   Lehn, J.-M. Pure Appl. Chem. **1994**, *66,* 1961.

10.   (a) Schmidtchen, F. P. J. Am. Chem. Soc.   **1986**, 108, 8249; (b) Schmidtchen, F. P. Tetrahedron Lett. **1989**, *30*, 4493.

11.   (a) Lehn, J.-M. Angew. Chem. Int. Ed. Engl.**1988**, *27,* 89; (b) Amit, A. G.; Mariuzza, R. A.; Phillips, S. E.; Poljak, R. J. Science **1986**, *233*, 747; (c) Geysen, H. M.; Tainer, J. A.; Rodda, S. J.; Manson, T. J.; Alexander, H.; Getzoff, E. D.; Lerner, R. A. Science, **1987**, *235*, 1184.

12.   Vogtle, F. Supramolecular chemistry, An Introduction; Wiley & Sons: Chichester, **1993**.

13. Chang, S.-K.; van Engen, D.; Fan, E.; Hamilton, A. D. J. Am. Chem. Soc. **1991**, 113, 7640.

14. (a)Iimori, T.; Still, W. C.; Rheingold, A. L. Staley, D. L. J. Am. Chem. Soc. **1989**, 111, 3439; (b) Cowart, M. D.; Sucholeiki, I.; Bukownik, R. R.; Wilcox, C. S. J. Am. Chem. Soc. **1988**, 110, 6204; (c)Dharanipragada, R.; Ferguson, S. B.; Diederich, F. J. Am. Chem. Soc. **1988**, 110, 1679.

15. Kitaigorodski, A. I. Molecular Crystals and Molecules, Academic Press: New York, **1973**.

16. (a) Beijer, F. H.; Kooijman, H.; Spek, A. L.; Sijbesma, R. P.; Meijer, E. W. Angew. Chem. Int. Ed. **1998**, 37, 75. (b) Djurdjevic, S.; Leigh, D. A.; McNab, H.; Parsons, S.; Teobaldi, G.; Zerbetto, F. J. Am. Chem. Soc. **2007**, *129*, 476.

17. Sessler, J. L.; Lawrence, C. M.; Jayawickramarajah, J. Chem. Soc. Rev. **2007**, *36*, 314.

18. Corey, E. J.; Becker, K. B.; Varma, R. K. J. Am. Chem. Soc. **1972**, *94,* 8616.

19. Crego, M.; Raposo, C.; Caballero, M. C.; García, E.; Saez, J. G.; Morán, J. R. Tetrahedron Lett.**1992**, 33, 7437.

20. Raposo, C.; martin, M.; Mussons, M. L.; Crego, M.; Anaya, J.; Caballero, M. C.; Moran, J. R. J. Chem. Soc. perkin trans 1 **1994**, 2113.

21. (a) Gust, D.; Moore, T. A. Topics, curr. Chem.**1991**, 159, 103; (b) Colquhoun, H. M.; Stoddart, J. F. Williams, D. J. Wolstenholme, J. B. Zarzycki, R. angew chem., int. ed. 1981, 20, 1051.

22. Yang, W.; Xia, P. F.; Wong, M. S. Org. Lett. **2010**, 12, 4018.

23. (a) de Gennes, P.G.; Prost, J. The Physics of Liquid Crystals. Oxford: Clarendon Press, **1993**. (b) Chandrasekhar, S. Liquid Crystals, 2nd Ed., Cambridge University Press, Cambridge, **1992**. (c) Sluckin, T. J.; Dunmur,

D. A.; Stegemeyer, H. Crystals That Flow - classic papers from the history of liquid crystals. Taylor & Francis, London, **2004**. (d) Martin, J. D.; Keary, C. L.; Thornton, T. A.; Novotnak, M. P.; Knutson, J. W.; Folmer J. C. W. Nature Materials **2006**, 5, 271. (e) Movahed, H. B.; Hidalgo, R. C.; Sullivan, D. E. Phys. Rev. **2006**, E73, 032701. (f) Stępień, M.; Donnio, B.; Sessler, J. L. Angew. Chem. Int. Ed. **2007**, 46, 1431.

24.   Wilson, A. G.; Brooke, O. G.; Lloyd, H. J.; Robinson, B. F. Br. Med. J. **1996**, *4*, 399.

25.   Armstrong, D. W.; Ward, T. J.; Armstrong, R. D.; Beesley, T. E. Science, **1986**, *232,* 1132.

26.   Bozdemir, O. A.; Sozmen, F.; Buyukcakir, O.; Guliyev, R.; Cakmak, Y.; Akkaya, E. U. Org. Lett. 2010, 12, 1400.

27.   Choi, M. G.; Cha, S.; Park, J. E.; Lee, H.; Jeon, H. L.; Chang, S.-K. Org. Lett. **2010**, 12, 1468.

28.   (a)Chakraborty, S.; Ravindran, V.; Nidheesh, P.V.; Rayalu, S. Chemistry Select, **2020,** *doi.org/10.1002/slct.202002113.* (b)Chakraborty,S.; Rayalu, S.; Spectrochim. Acta - Part A Mol. Biomol. Spectrosc.,2020, *https://doi.org/10.1016/j.saa.2020.118915* (c)Chakraborty, S.; Mandal, M.; Rayalu, S.; Inorgnica Chimica Acta, **2020,** https://doi.org/10.1016/j.inoche.2019.107693. (d)He, G.; Zhao, X.; Zhang, X.; Fan, H.; Wu, S.; Li, H.; He, C.; Duan, C. New. J. Chem. **2010**, *34,* 105(e) Goswami, S.; Chakraborty, S.; Paul, S.; Halder, S.; Panja, S.; Mukhopadhyay, S. K. Org. Biomol. Chem. **2014,** *12* (19), 3037. (With inside cover page image).(f) Goswami, S.; Chakraborty S.;, Adak, M. K.; Halder, S, Quah, C. K.; Fun, H. K.; Pakhira, B.; Sarkar, S. New J. Chem. 2014, 38, 6230. (g)Goswami.S.; Chakraborty, S.; Das, A. K.; Manna, A.; Bhattacharyya, A.; Quah, C. K.; Fun, H. K. Rsc Adv. **2014,** *40*, 20616. (h)

Goswami, S.; Chakraborty, S.; Paul, S.; Halder, S.; Maity, A. C. Tetrahedron Lett. **2013**, *54,* 5075. Zapata, F.; Caballero, A.; Espinosa, A.; Tarraga, A.; Molina, P. Org. Lett. **2008,** *10,* 41.

29. Sheng, R.; Wang, P.; Gao, Y.; Wu, Y.; Liu, W.; Ma, J.; Li, H.; Wu, S. *Org. Lett.* **2008,** *10,* 5015.

30. (a) Miller, J. N.; Birch, D. J. S.; **1997**, 4th International Conference on Methods and Applications of Fluorescence Spectroscopy, J. Fluoresc. 7(1): 1S-246S; (b) Thompson, R. B. **1997**, Advance in Fluorescence Sensing Technology III, Proc. SPIE **2980**; (c) Lakowicz, J. R.; **1995**, Advance in Fluorescence Sensing Technology II, Proc. SPIE **2388**.

31. Zhou, Y.; zhu, C.-Y.; Gao, X.-S.; You, X.-Y.; Yao, C. Org. Lett. **2010**, *12,* 2566.

32. (a) de Silva, A. P.; Gunaratne, H. Q. N.; Habib-Jiwan, J.-L.; McCoy, C. P.; Rice, T. E.; Soumillion, J.-P. Angew. Chem. Int. Ed. Engl.**1995**, 34, 1728; (b) Bryan, A. J.; de Silva, A. P.; De Silva, S. A.; Rupasinghe, R. A. D. D.; Sandanayake, K. R. S. S. *biosensors,* **1989**, *4,* 169.

33. Kelly, T. R.; Bridger, G. J.; Zhao, C. J. Am. Chem. Soc. **1990**, 112, 8024.

34. Lehn, J.-M. *Angew. Chem. Int. Ed. Engl.***1988**, *27,* 89.

35. (a) Kubik, T.; Bogunia-Kubik, K.; Sugisaka, M. Curr. Pharm. Biotechnol. **2005**, *6,* 17. (b) Cavalcanti, A.; Freitas, R. A. Jr. IEEE Trans Nanobioscience **2005**, 4, 133. (c) Cavalcanti, A.; Shirinzadeh, B.; Freitas, R. A. Jr.; Kretly, L. C. Recent Patents on Nanotechnology. **2007**, 1, 1. (d) Sada, K.; Takeuchi, M.; Fujita, N.; Numata, M.; Shinkai, S. Chem. Soc. Rev. **2007**, 36, 415.

36. (a) Sessler, J. L.; Eller, L. R.; Cho, W.-S.; Nicolaou, S.; Aguilar, A.; Lee, J. T.; Lynch, V. M.; Magda, D. J. Angew. Chem. Int. Ed. **2005**, 44, 5989. (b) Naumovski, L.; Sirisawad, M.; Lecane, P.; Chen, J.; Ramos, J.; Wang,

Z.; Cortez, C.; Magda, D.; Thiemann, P.; Boswell, G.; Miles, D.; Cho, D. G.; Sessler, J. L.; Miller, R. Mol Cancer Ther. **2006**, 11, 2798.

(c) Pedersen, C. J. Angew. Chem. Int. Ed. **1988**, 27, 1021.

(d) Schmidtchen, F. P. J. Am. Chem. Soc. **1986**, 108, 8249. (e) Schmidt chen, F. P. Tetrahedron Lett. **1989**, 30, 4493. (f) Lehn, J. M. Angew. Chem. Int. Ed. **1988**, 27, 89. (g) Iimori, T.; Still, W. C.; Rheingold, A. L. Staley, D. L. J. Am. Chem. Soc. **1989**, 111, 3439. (h)Scheiner, S. Molecular Interactions. From Van der Waals to Strongly Bound Complexes, Wiley, Chichester, **1997**.

Publisher: Eliva Press SRL

Email: info@elivapress.com